AF690514

ACKNOWLEDGEMENT

ILLUSTRATED BY:

M.Bilal

From

(Artigence)

You are the greatest superhero that has ever lived. Because your body is the most amazing creation in all the world and capable of performing unlimited super feats.

Your body is made up of super wonderful things which help you eat, talk, see, touch, feel, smell and watch your favorite movies, do homework and most importantly, play!

There are several body parts which help us in our daily lives. Starting with our faces; each one is unique. No two faces are exactly alike.Your unique face makes you special and super amazing.

Your face is so amazing that it has the ability to create over 10,000 different expressions. That sounds like the power of a superhero.

Your eyes are also super amazing for they can see a candle flame up to 2 miles away. Your eyes can blink 15 - 20 times a minute and up to 1,200 blinks within an hour and 28,000 blinks within a single day!

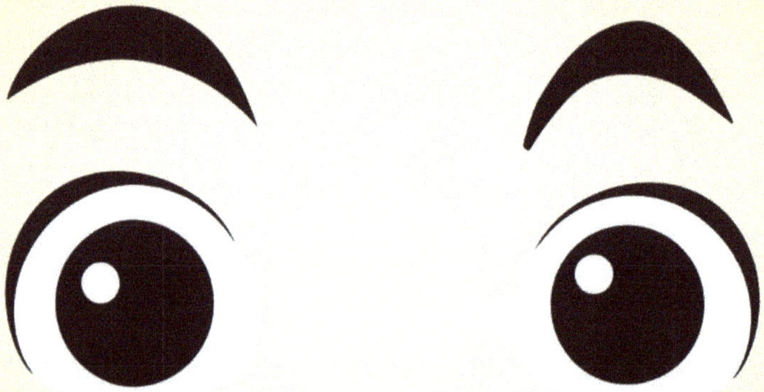

SUPER EYES

You also have soft cheeks which puff up and lips which grow wide when we smile! Your lips can create 19 different types of amazing smiles. Only a superhero can do that!

You also have wonderful ears which help you listen. Your ears are so super amazing that they can even clean themselves using hairs that remove dust, particles and wax. Your ears can also hear sounds as far as 2 miles away. Yes, you do have super hearing!

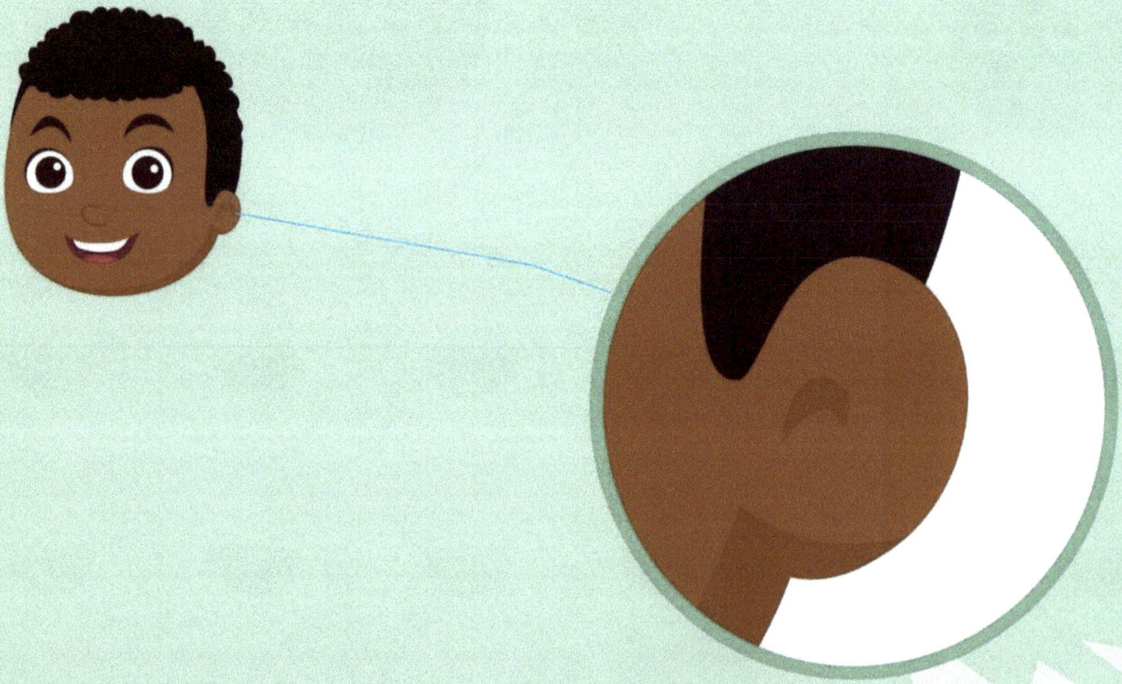

SUPER EAR

"Your nose, like your ears, is quite amazing too for it can detect 1 trillion different smells from all around. Super smelling is just one of your many super gifts!

Your mouth is amazing because it helps you to talk. Unlike all other creatures you have the power to form sentences. About 100 muscles located within the chest, neck and jaw work together to allow you the ability to form a single phase like, "Hello, how are you?" Because of these muscles you have the ability to speak many languages. Some people can speak up to 40 languages from countries all around the world. Do you have this superpower too?

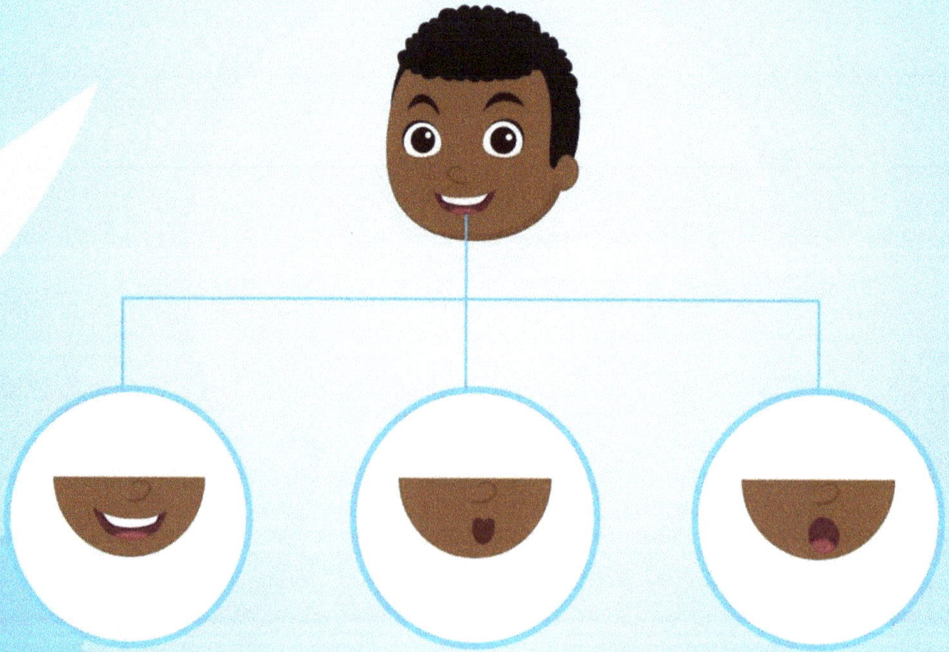

Inside your mouth you have the superpower to create saliva to help break down and digest food. Your mouth is so amazing that it will create 37,854 liters of saliva in a lifetime - that's enough to fill up 2 swimming pools to the brim. Now that's super amazing!

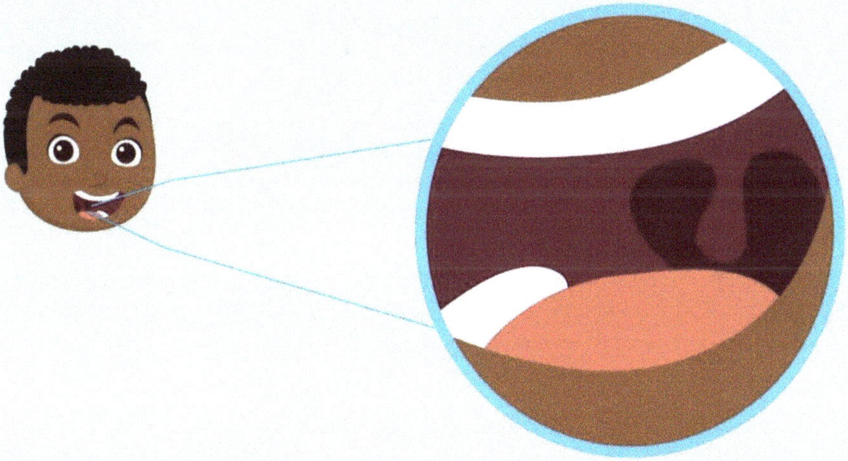

SUPER SALIVA

Your Teeth is also amazing and allow you to chew your food. Your teeth are as strong as a Shark's teeth and your bite can exert a force of 200 pounds of pressure when you chomp down. Your teeth are so super awesome that they have been known to last for a hundred years! How super can you be?

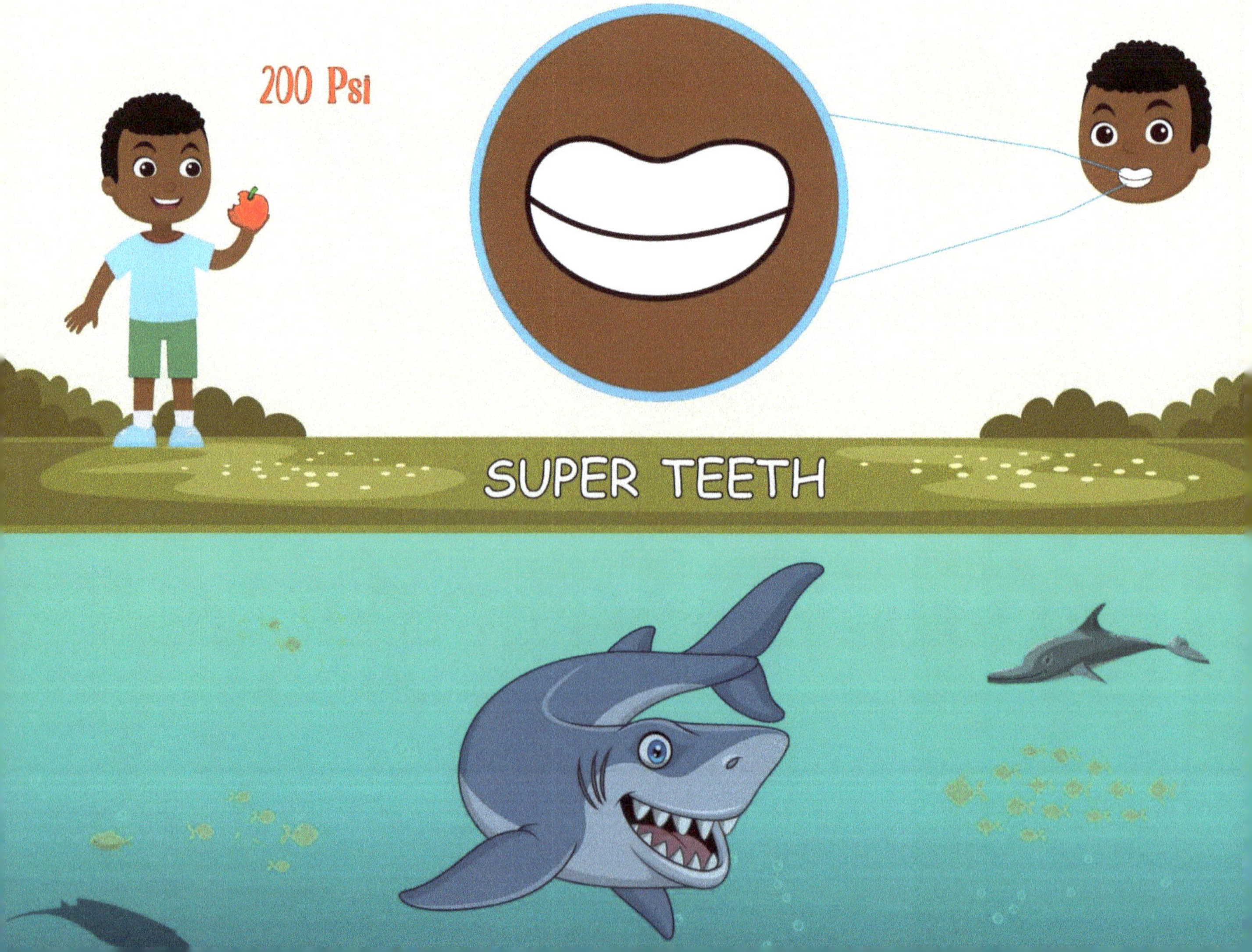

Your tongue is another amazing part of your body because it is covered with 8000 taste buds to help you enjoy all that you eat. Now that's a lot of taste buds.

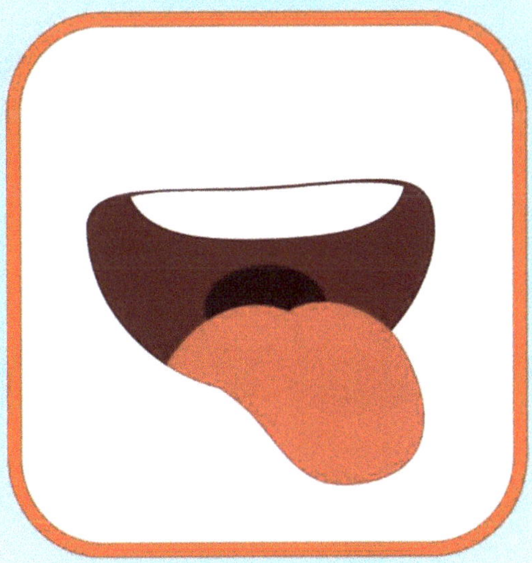

SUPER TONGUE

Your hair is amazing too and so thin that it can be cut and styled in different ways. But your hair is also super strong and a full head of hair can lift 2 Elephants or up to 2-tons of weight! That is super amazing!

We use our hands the most so our hands are really useful and strong too. Your hands are the most amazing from all other creatures because you have a thumb. Your thumb allows you to perform thousands of different activities that other creatures cannot do like playing the piano, playing video games or typing on the computer. Only a superhuman can do all of this!

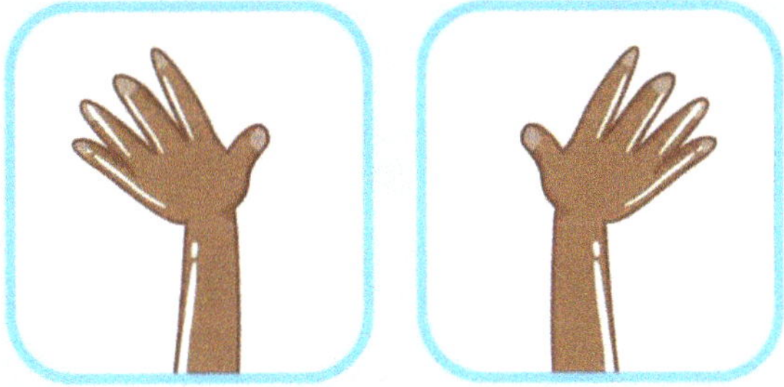

SUPER HANDS

Your legs are also super amazing and can allow you to jump up to 20 inches into the air and run up to 2 miles per hour. Now that is super amazing.

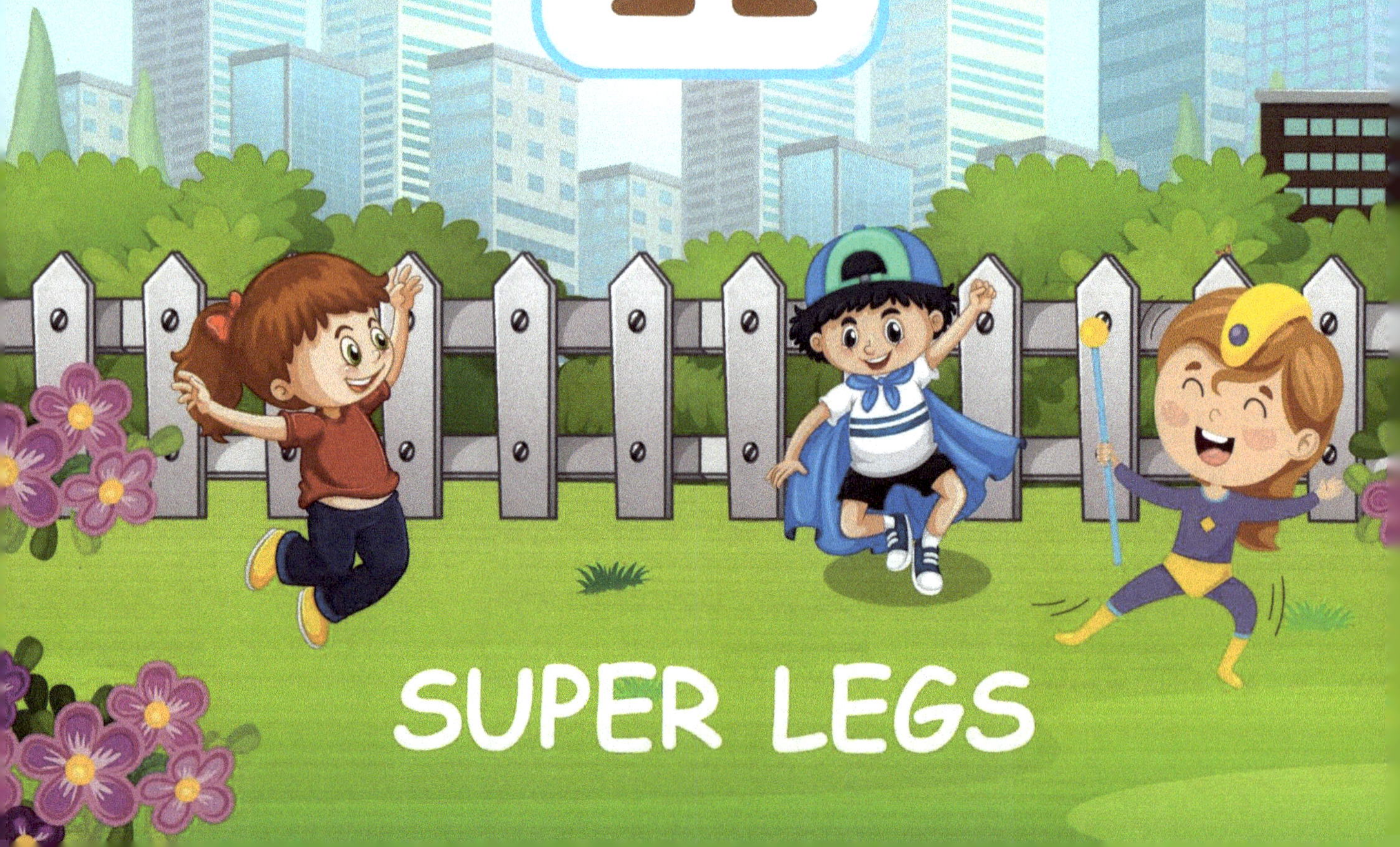

SUPER LEGS

Your stomach digests your food when you eat and then sends the nutrients to various parts of the body. Inside the stomach are fluids called acids that help digest your food. The acids in the stomach are so awesome that they can even dissolve and break apart metals like steel and copper. Your superpowers are amazing!

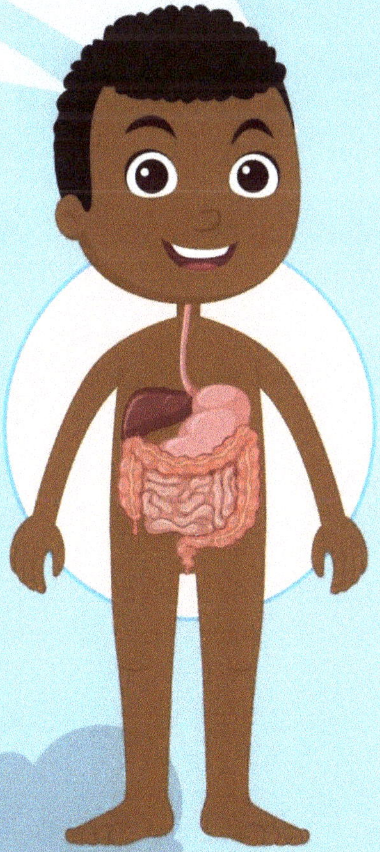

SUPER STOMACH

Your liver helps to digest your food. Your liver is so amazing because it is an organ that can completely regenerate itself all by itself. Now, do you know anything else in the world that has the power to regenerate?

SUPER LIVER

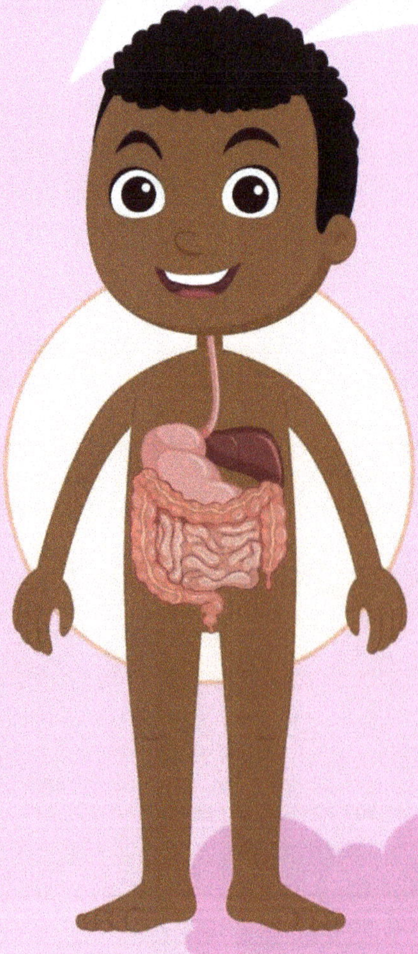

Almost all the food you eat gets broken down inside the small intestine and the nutrients are absorbed into the body. But if you tried to measure the small intestine you would need a very long ruler for it is 22 feet long, yet it can still all fit inside your stomach. Now that is super amazing!

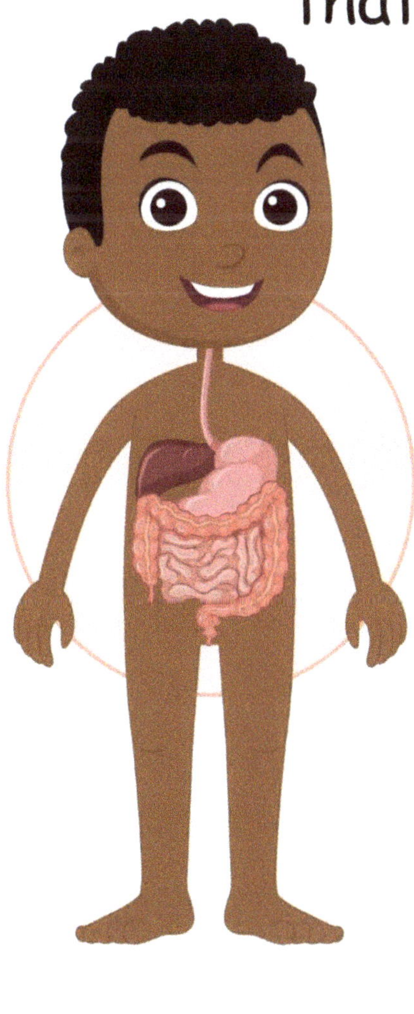

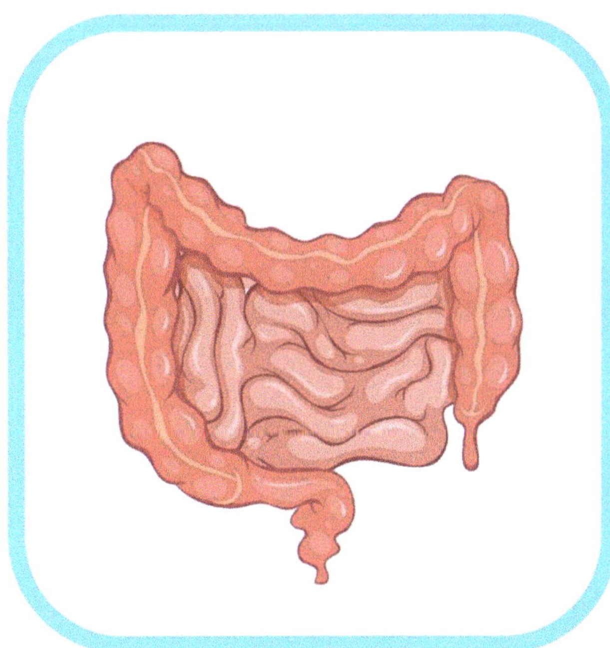

SUPER INTESTINE

Your beautiful skin which comes in so many different shades is so amazing that it will completely renew itself every 28 days. The skin is made up of tiny cells. When it's ready to renew itself, the skin will shed up to 40,000 cells per minute. Only the greatest of superheroes can do this.

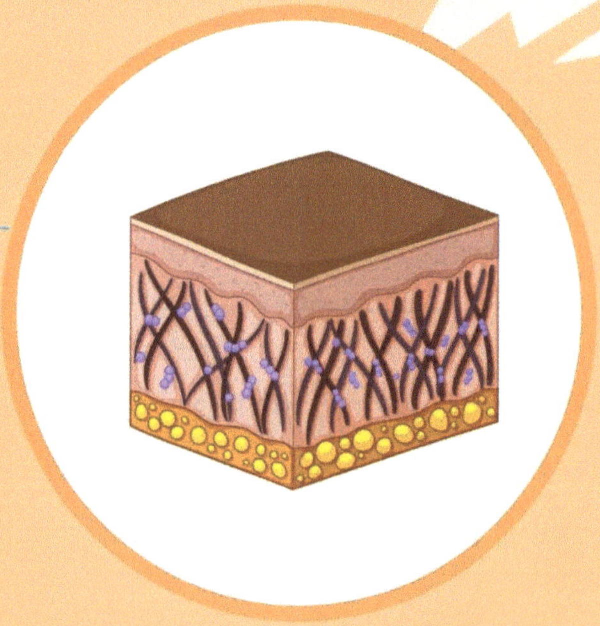

SUPER SKIN

You have 206 bones in your body and they are all amazing! Your bones are so awesome that they are stronger than steel. They are so strong that they can bear a heavy load up to 19,000 pounds per square inch. That is like lifting 5 pick-up trucks all at once with little effort. You are simply super amazing!

SUPER BONES

In addition to your great bone strength, your skeleton is so awesome that every 10 years it completely replaces itself. That means your body will create a new set of bones every ten years for the rest of your life!

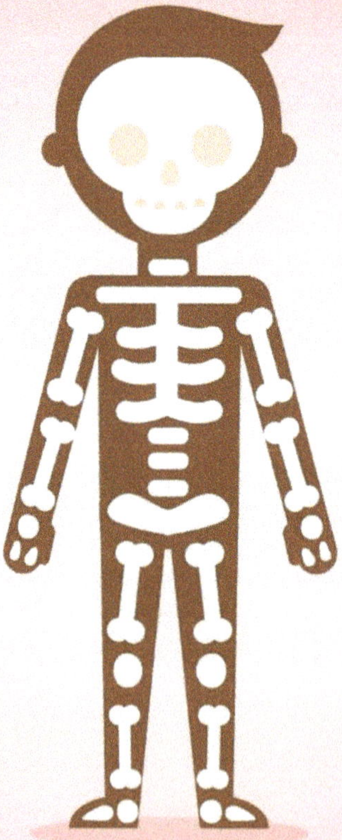

SUPER SKELETON

The greatest computer in the world is a supercomputer called the Human Brain. Your brain is so amazing that it can perform a billion, billion calculations per second. Nothing in the world can do that. This means your brain is the most powerful of all the computers on the planet. This is your greatest superhero power. Now go and amaze the world!

SUPER BRAIN

Your body is special and the most amazing creation in the universe. Take care of your body and do amazing things with all of your great superpowers!

You can achieve anything that you want!

THE END

REFRENCES

Buffoli, Barbara; Rinaldi, Fabio; Labanca, Mauro; Sorbellini, Elisabetta; Trink, Anna; Guanziroli, Elena; Rezzani, Rita; Rodella, Luigi F. (2014). The human hair: from anatomy to physiology. International Journal of Dermatology.

Clemente, Carmine (1987). Anatomy, a regional atlas of the human body. Baltimore: Urban & Schwarzenberg. ISBN 978-0-8067-0323-7.

D Warfield. 1973. The study of hearing in animals. In: W Gay, ed., Methods of Animal Experimentation, IV. Academic Press, London, pp 43-143.

A Global Perspective on Bilingualism and Bilingual Education (1999), G. Richard Tucker, Carnegie Mellon University" (PDF). cal.org. Retrieved 8 May 2018.

The Cells in Your Body. Science NetLinks.

Anatomy of the Human Body. Henry Gray. 20th edition.

ABOUT THE AUTHOR

The author is an educator and a passionate lover of children. He received his Bachelor of Arts in Biology from Edgewood College, Madison in Wisconsin and a second B.A. in Healthcare Management from Southern Illinois University Carbondale where he also completed his M.A in Engineering. He has a wealth of valuable experience dealing with children of varying age ranges. The author does not only love children, but he understands that when they have the best form of education, it has a far-reaching effect on the wider society. As a result, he is on a mission to make sure that children learn as much as they can.

www.ingramcontent.com/pod-product-compliance
Lightning Source LLC
LaVergne TN
LVHW070900210926
848482LV00016B/983

* 9 7 9 8 9 9 7 3 6 6 1 4 8 *